Carbon

and the Elements of Group 14

Nigel Saunders

Heinemann Library
Chicago, Illinois

Design: David Poole and Tinstar Design
    Limited (www.tinstar.co.uk)
Illustrations: Geoff Ward and Paul Fellows
Photo Research: Rosie Garai
Originated by Blenheim Colour Ltd.
Printed in China by South China Printing Company.
07 06 05 04 03
10 9 8 7 6 5 4 3 2 1
Library of Congress Cataloging-in-Publication Data

Saunders, N. (Nigel)
  Carbon and the Group 14 elements / Nigel Saunders.
      v. cm. -- (The periodic table)
Includes bibliographical references and index.
Contents: Elements and atomic structure -- The periodic table, Group 14
and carbon -- Carbon -- Silicon -- Germanium -- Tin -- Lead -- Useful
information about the Group 14 elements.
  ISBN 1-40340-873-4 (lib. bdg.) -- ISBN 1-40343-516-2 (pbk.)
  1. Carbon--Juvenile literature. 2. Group 14 elements--Juvenile
literature. [1. Carbon. 2. Group 14 elements.] I. Title. II. Series.
  QD181.C1 S26 2003
  546'.68--dc21
                            2002154192

Acknowledgments
The author and publishers are grateful to the following for permission to reproduce copyright material:
p. **4** Robin Smith Stone/Getty Images; p.**8** Peter Menzel/Science Photo Library; p. **9** Charles Falco/Science Photo Library; p. **10** Stuart Westmorland/Corbis; p. **11** Vanni Archive/Corbis; pp. **12**, **13** Martin Bond/Science Photo Library; p. **15** Rick Gayle/Corbis; p. **17** Andrew McClenaghan/Science Photo Library; p. **19** Lee Snider/Corbis; p. **21** Kaj R. Svensson/Science Photo Library; p. **22** Martin Bond/Science Photo Library; pp. **25**, **32** Trevor Clifford/Science Photo Library; p. **26** Matt Meadows/Science Photo Library; p. **29** Adam Jones/Oxford Scientific Films; p. **30** Astrid & Hans Frieder Michler/Science Photo Library; p. **37** Kazuyoshi Nomachi/Science Photo Library; **p. 38** B. Kramer/Custom Medical Stock Photo/Science Photo Library; p. **40** Adam Woolfitt/Corbis; p. **43** David Taylor/Science Photo Library; p. **44** Novosti/Science Photo Library; pp. **47**, **49** Arnold Fisher/Science Photo Library; p. **48** Adam Woolfitt/Corbis; p. **50** Bob Krist/Corbis; p. **52** Owen Franken/Corbis; pp. **54**, **55** Oscar Burriel/Science Photo Library; p. **57** Geoff Tompkinson/Science Photo Library.

Cover photograph of diamonds reproduced with permission of Getty Images.

Special thanks to Theodore Dolter for his review of this book.

Every effort has been made to contact copyright holders of any material reproduced in this book. Any omissions will be rectified in subsequent printings if notice is given to the publishers.

Words appearing in bold, **like this,** are explained in the Glossary.

Contents

DUFFIELD BRANCH
2507 W. Grand Blvd.
Detroit, MI 48208

CHAPTER 1

Elements and Atomic Structure

Have you ever wondered how many different substances there are in the world? If you look around, you'll see metals, plastics, water, and lots of other solids and liquids. Although you can't see the gases in the air, you know they are there. There are many other gases, too. Just how many different substances are there? Incredibly, over 19 million different substances have been discovered, named, and cataloged. Around 4,000 substances are added to the list each day. All of these substances are made from just a few simple building blocks called **elements.**

Elements

There are about 90 naturally occurring elements and a few artificial ones, including element number 114 in group 14. Elements are materials that cannot be broken down into simpler substances using chemical **reactions.** Close to three-quarters of the elements are metals, such as tin and lead, and most of the rest are nonmetals, such as carbon and oxygen. Some elements, like germanium and silicon, are called metalloids because they have some of the properties of metals and some of the properties of nonmetals.

This thrilling roller coaster and its riders are made from some of the millions of chemicals in the world. ▶

Compounds

Elements can join together in chemical reactions to make **compounds.** For example, lead and oxygen react together to make lead oxide, and carbon and oxygen react together to make carbon dioxide. Nearly all of the millions of different substances in the world are compounds, made up of two or more elements chemically joined together.

Atoms

Every substance is made up of tiny particles called **atoms.** An element is made up of just one type of atom, and a compound is made up of two or more types of atom joined together. Atoms are far too small to be seen, even with a light microscope. If you could line up carbon atoms side by side along a fifteen-centimeter (six-inch) ruler, you would need a billion of them!

Atoms themselves are made up of even tinier particles called **protons, neutrons,** and **electrons.** At the center of each atom there is a **nucleus** made up of protons and neutrons. The electrons are arranged in different energy levels, or shells, around the nucleus. Most of an atom is actually empty space—if an atom were blown up to the same size as an Olympic running track, its nucleus would be about the size of a pea! The electrons, and how they are arranged, are responsible for the ways in which each element can react.

Elements and groups

Different elements react with other substances in different ways. When scientists first began to study chemical reactions, this made it difficult for them to make sense of the reactions they observed. In 1869, a Russian chemist named Dimitri Mendeleev put each element into one of eight **groups** in a table. Each group contained elements with similar chemical properties. This made it much easier for chemists to figure out what to expect when they reacted elements with each other. You can find the modern equivalent, the **periodic table,** on the next page.

The Periodic Table, Group 14, and Carbon

Chemists built on Mendeleev's work and eventually produced the modern **periodic table**, shown below. Each row in the table is called a **period**. The elements in a period are arranged in order of increasing **atomic number** (the atomic number is the number of **protons** in the **nucleus**). Each column in the table is called a **group**. The **elements** in each group have similar chemical properties to each other. For example, the elements in group I are very reactive, soft metals, while the elements in group 0 (zero) are very unreactive gases. It is called the periodic table because these different chemical properties occur regularly, or periodically. The elements in each group also have the same number of **electrons** in their outer shell. The elements in group IV all have four electrons in their outer shell.

▼ *This is the periodic table of the elements. The metals are on the left, and the nonmetals are on the right. Group IV contains a nonmetal (carbon), two metalloids (silicon and germanium), and three metals (tin, lead, and ununquadium).*

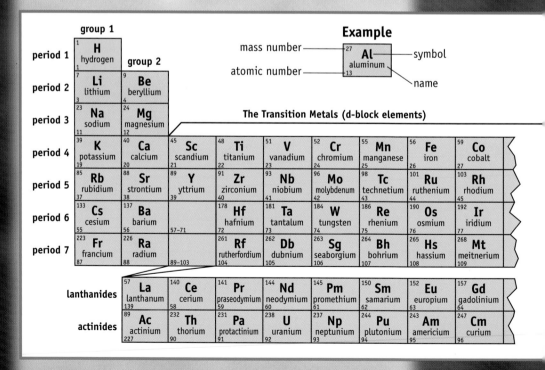

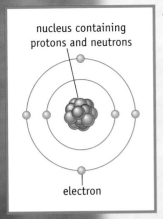

nucleus containing
protons and neutrons

electron

◀ *This model shows an atom of carbon. Every element has a different number of protons (atomic number). A carbon atom contains six protons and six neutrons. Its electrons are arranged in two shells around the nucleus.*

As you go down each group, the chemical properties of the elements change gradually. Carbon, at the top of group 14, is a nonmetal with a high melting point. Silicon and germanium, in the middle of the group, are metalloids, and tin and lead at the bottom are metals with low melting points. There is also an artificial element in group 14, below lead. It is temporarily called ununquadium (pronounced "yoon-oon-kwad-ee-um"), and very few **atoms** of it have been made. We do not know very much about it because its atoms give off radiation and break down to form other elements within seconds of being made. However, chemists are certain that it is a metal because the elements immediately above it in group 14 are both metals.

Group 14 and carbon

In this book, you are going to find out all about carbon and the other elements in group 14, as well as the **compounds** they make and many of their uses.

Key

	metals
	metalloids
	nonmetals

		group 13	group 14	group 15	group 16	group 17	group 18	
							4 **He** helium 2	period 1
		11 **B** boron 5	12 **C** carbon 6	14 **N** nitrogen 7	16 **O** oxygen 8	19 **F** fluorine 9	20 **Ne** neon 10	period 2
		27 **Al** aluminum 13	28 **Si** silicon 14	31 **P** phosphorus 15	32 **S** sulfur 16	35 **Cl** chlorine 17	40 **Ar** argon 18	period 3
59 **Ni** nickel 28	64 **Cu** copper 29	65 **Zn** zinc 30	70 **Ga** gallium 31	73 **Ge** germanium 32	75 **As** arsenic 33	79 **Se** selenium 34	80 **Br** bromine 35	84 **Kr** krypton 36	period 4
106 **Pd** palladium 46	108 **Ag** silver 47	112 **Cd** cadmium 48	115 **In** indium 49	119 **Sn** tin 50	122 **Sb** antimony 51	128 **Te** tellurium 52	127 **I** iodine 53	131 **Xe** xenon 54	period 5
195 **Pt** platinum 78	197 **Au** gold 79	201 **Hg** mercury 80	204 **Tl** thallium 81	207 **Pb** lead 82	209 **Bi** bismuth 83	209 **Po** polonium 84	210 **At** astatine 85	222 **Rn** radon 86	period 6
269 **Ds** darmstadtium 110	272 **Uuu** unununium 111	269 **Uub** ununbium 112		289 **Uuq** ununquadium 114		292 **Uuh** ununhexium 116			period 7

159 **Tb** terbium 65	163 **Dy** dysprosium 66	165 **Ho** holmium 67	167 **Er** erbium 68	169 **Tm** thulium 69	173 **Yb** ytterbium 70	175 **Lu** lutetium 71	⎫ f block
247 **Bk** berkelium 97	251 **Cf** californium 98	252 **Es** einsteinium 99	257 **Fm** fermium 100	258 **Md** mendelevium 101	259 **No** nobelium 102	262 **Lr** lawrencium 103	⎭

Elements of Group 14

There are six **elements** in group 14: carbon, silicon, germanium, tin, lead, and ununquadium. They are all solids at room temperature. Carbon is a nonmetal, and tin, lead, and ununquadium are metals. Silicon and germanium are metalloids, meaning that their properties are between those of metals and nonmetals.

12	
C	**carbon**
carbon	symbol: C • atomic number: 6 • nonmetal
6	

Coal, charcoal, and soot are almost pure carbon, so it is likely that people have known about carbon for thousands of years. The name carbon comes from the Latin word meaning charcoal. However, carbon was not clearly recognized as an element until the 18th century.

There are different forms of carbon that look very different from each other. These different forms are called **allotropes** of carbon. Graphite and diamond are the best-known allotropes of carbon. Graphite is the shiny black solid used in pencil "lead," and diamond is the clear, colorless solid often seen in jewelry. These allotropes exist because of the different ways carbon **atoms** can join together. Carbon also exists as **amorphous** carbon (found in soot), as microscopic hollow tubes called nanotubes, and as spherical molecules called fullerenes.

Carbon can form a vast number of **compounds**, and it is found almost everywhere on Earth. Carbon dioxide makes up 0.035 percent of the atmosphere and it is dissolved in rivers, lakes, and seas. Many **minerals** and rocks contain carbon compounds. Natural gas and crude oil contain compounds of hydrogen and carbon called **hydrocarbons.** Chains and rings of carbon

Crude oil gushes out of a damaged oil well in ▶
Kuwait. Oil contains a complex mixture of many
different carbon compounds.

atoms are the vital component in the complex molecules of life itself, and all living things contain carbon. There is a lot of carbon in your body—over 20 percent of your mass is due to carbon!

28 Si silicon 14	**silicon** *symbol: Si • atomic number: 14 • metalloid*

The Swedish chemist Jöns Berzelius, who introduced many of the chemical symbols we use today, discovered silicon in 1823. There are two allotropes of silicon: a dark red-brown powder called amorphous silicon, and a shiny gray-black solid called crystalline silicon.

Silicon is the second most abundant element in Earth's crust, but it is usually found combined with the most abundant element, oxygen, in various silicon dioxides. Sand is the best-known silicon dioxide, but there are many others as well, such as quartz and flint. The name silicon comes from the Latin word for flint. Minerals such as granite and asbestos also contain silicon, and tiny living things called diatoms build their cell walls using silicon compounds.

Silicon and its compounds are extremely useful to us. Silicon combines with tiny amounts of other elements to make an important **semiconductor** material, used widely in electronic devices. Silicon compounds are used in glass, concrete, brick, pottery, and silicone sealants.

▼ *This computer microchip contains very pure crystalline silicon with tiny amounts of other elements as well.*

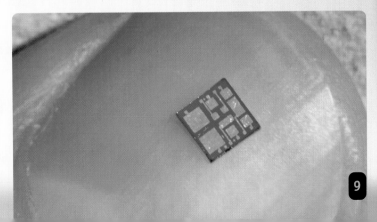

9

More Elements of Group 14

73	Ge
germanium
32	

germanium

symbol: Ge • atomic number: 32 • metalloid

Clemens Winkler, a German chemist, discovered germanium in 1886. The name germanium comes from the Latin word for Germany.

Germanium is a **brittle** gray-white crystalline solid found in a number of **minerals.** These include argyrodite (from which it was first isolated) and germanite, which contains 8 percent germanium.

Germanium is an important **semiconductor** material used in all sorts of electronic devices. Telephone and cable television networks use optical fibers made from glass containing germanium to carry their signals. Germanium is also used in fluorescent lights and in high-quality lenses for microscopes and cameras.

119	Sn
tin
50	

tin

symbol: Sn • atomic number: 50 • metal

Tin is an Anglo-Saxon word, but its symbol comes from the Latin word for tin (*stannum*). Tin metal cannot be found in its native form as a free **element,** but it has been known for thousands of years. It is clear that tin has been in use at least since the Bronze Age (about 4,000 years ago), because bronze is an **alloy** that is made from tin and copper. Cassiterite, or tin (IV) oxide, is the most important kind of tin **ore.**

When barnacles and other sea creatures live on the hulls of ships, the ships have to use more fuel. Paints containing tin compounds stop these creatures from growing. However, on sunken wrecks they can grow freely.

There are two **allotropes** of tin, and they can be converted into each other by heating or cooling. Ordinary tin is a **malleable** silvery-white solid. When this is cooled below 55.76 °F (13.2 °C), it slowly changes into a brittle gray solid that turns back into white tin when it is warmed up again. Tin does not react with water, but it will react with strong acids and alkalis, and with oxygen when heated in air.

Tin is important in making glass, but its most familiar use is in cans for storing food. It is also used in alloys such as solder, bronze, and pewter. Most tin **compounds** are toxic, and some are used to prevent barnacles and other sea creatures from growing on the hulls of ships.

207 **Pb** lead 82	**lead** *symbol: Pb • atomic number: 82 • metal*

Lead is an Anglo-Saxon word, but its symbol comes from the Latin word for lead (*plumbum*). Lead metal is only rarely found in its native form as a free element, but like tin, it has been known for thousands of years. Galena, or lead (II) sulfide, is the most important lead ore. Lead is a very soft, shiny, blue-white metal. It is fairly unreactive because it is protected by a thin layer of gray lead (II) oxide.

Lead has many uses, including radiation shielding, car batteries, bullets, glass, and **insecticides.** Many lead compounds are brightly colored and are used in **pigments** and paints. However, lead is poisonous, so it has to be handled carefully.

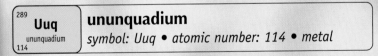

289 **Uuq** ununquadium 114	**ununquadium** *symbol: Uuq • atomic number: 114 • metal*

Only a few **atoms** of this metal have been made. Its name means "one-one-four" and is temporary until scientists know more about the element. Ununquadium was first produced in 1998 by smashing high-speed calcium atoms into plutonium atoms in a machine called a particle accelerator.

Carbon

Three **allotropes** of carbon occur naturally: diamond, graphite, and **amorphous** carbon (found in coal, charcoal, and soot). These have been known for thousands of years, but it was not until the 18th century that people realized they were all forms of the same **element.**

In 1772 a French chemist named Antoine Lavoisier showed that diamonds are made of carbon. He weighed samples of amorphous carbon and diamond, carefully burned them, and then weighed the **products** left behind. Lavoisier found that carbon and diamond both produced the same amount of carbon dioxide when they were burned. He realized that because carbon dioxide was the only product in both cases, diamond must be made of carbon. A Swedish chemist, Carl Scheele, did a similar experiment in 1779 to show that graphite is also made of carbon.

▲ *Lavoisier is considered one of the greatest experimentalists in chemistry.*

When carbon burns, it reacts with oxygen from the air and produces carbon dioxide. The equation for this is:

$$\text{carbon} + \text{oxygen} \rightarrow \text{carbon dioxide}$$
$$C(s) + O_2(g) \rightarrow CO_2(g)$$

Amorphous carbon

Amorphous carbon is a black solid. Soot and charcoal are forms of amorphous carbon, and it is also found in coal. It is formed when substances containing carbon are burned in a limited supply of oxygen. For example, charcoal can be made by burning wood, coconut shells, or animal bones.

If charcoal is heated with steam to about 1,832 °F (1,000 °C) while keeping oxygen away, it turns into activated charcoal. This is a very pure form of carbon that is porous, like a sponge. Chemical **reactions** involving a solid can happen only at its surface, so a solid with a big surface area can react very quickly. The tiny pores in activated charcoal give it a huge surface area, and 0.035 ounce (1 gram) of it can have the surface area of four tennis courts! It is very good at absorbing other chemicals, so it is used to purify food and water. Range hoods often have filters containing activated charcoal to stop cooking smells from escaping into the kitchen.

Amorphous carbon is molded into shape by putting it under high pressure and used to make the cores of batteries. However, about 90 percent of it is used in the rubber industry, mostly for tires. Carbon gives tires their familiar black color, and it also strengthens the rubber so that the tire is not worn away too quickly. Plastics often have amorphous carbon added to reinforce them and to protect them from damaging ultraviolet light from the Sun. You will often see this black plastic used for drainpipes and electrical cables.

Amorphous carbon is used in black paints and in the black inks used in inkjet printers. The toners used in photocopiers and laser printers contain very finely powdered amorphous carbon. As the paper passes through the machine, carbon is baked onto the paper to form the image. Each particle of carbon in the toner can be as small as 0.00039 inch (0.01 millimeter) in diameter. This makes very detailed images possible.

The rubber in tires contains carbon to strengthen it.

Diamond

Pure diamond is a clear and colorless solid, but most diamonds have a faint yellow color because they contain traces of other **elements.** Diamond is the hardest natural substance known, though it is also very **brittle** and can be smashed with a hammer. It does not conduct electricity, but it is an excellent conductor of heat.

The structure of diamond

Nonmetals like carbon join on to other **atoms** using chemical **bonds** called covalent bonds. These bonds form when two atoms share a pair of **electrons.** Every carbon atom can make four covalent bonds. This means that a carbon atom can join to four other atoms using four single covalent bonds, as in methane, CH_4, or to two other atoms using double covalent bonds, as in carbon dioxide, CO_2, or in other combinations. In diamond, every carbon atom is joined to four other carbon atoms using single covalent bonds.

A crystal of diamond is a single giant molecule, called a macromolecule. A molecule of diamond contains a huge number of atoms and bonds. In fact, a diamond only 0.2 inch (5 millimeters) in diameter contains billions of carbon atoms. Diamonds are very hard because each bond is strong and there are very many of them. It also has a very high melting point, 6,422 °F (3,550 °C), because a lot of energy is needed to break all the bonds.

Each carbon atom can make four chemical bonds with other atoms. The bonds are arranged so that they point to the four points of an imaginary three-sided pyramid.

bonds

atom

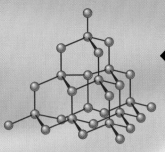

In the structure of diamond, each carbon atom (except the ones at the surface) is joined to four other carbon atoms to make a giant molecule.

Mining and manufacture

Natural diamonds formed underground millions of years ago in molten rock, called magma, under huge pressures and temperatures. As the magma flowed towards the surface through volcanic pipes, it cooled and solidified to form a blue rock called kimberlite. Miners dig this out of the ground to get at the diamonds. Diamonds are found all over the world, but most diamond mines are in South Africa and the Democratic Republic of Congo.

Artificial diamonds can be made for industrial use. Small diamonds just one-eighth inch (three millimeters) across are made by dissolving graphite in molten nickel, then heating and squashing it for several hours. Thin films of diamond are made on the surface of other substances by heating carbon to a high temperature at low pressure. This causes the carbon to turn into a vapor that then reforms as a film of diamond on the surface.

The uses of diamond

Clear diamonds with a good color are called gem diamonds. Before they can be used for jewelry, they need to be cut to size and shape by highly skilled diamond cutters so that they sparkle in the light. Only about a quarter of mined diamonds are good enough for jewelry—the rest are used with artificial diamonds for industry.

Industrial diamonds are used as edges on cutting tools for drilling oil wells. Diamond films are used in electronic devices, where they conduct heat away from computer chips and other components.

Diamonds like these can be used in jewelry. They are often set in precious metals to show off their high quality.

Graphite

Graphite is a **brittle,** shiny black solid. It has a soft, slippery feel and leaves a black mark when rubbed on paper, meaning that it can be used in pencils. Like diamond, it does not dissolve in water and it has a high melting point, but unlike diamond it conducts electricity.

The structure of graphite

Graphite consists of layers of carbon **atoms** stacked on top of one another, with very weak **bonds** between them. These weak bonds allow the layers to slide over each other very easily, causing graphite to be soft and slippery.

In the structure of graphite, each carbon atom in a layer is joined to three other carbon atoms to make hexagonal rings. There are very weak bonds between the layers.

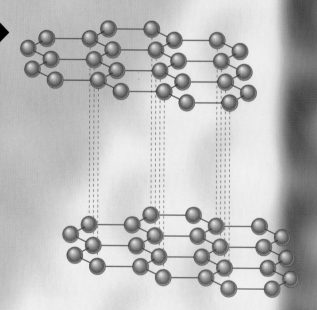

The carbon atoms in a layer of graphite are joined using single covalent bonds. The rings of six atoms that they form are joined so that each layer looks like a honeycomb or a piece of chicken wire. There are "spare" **electrons** in every layer, because each carbon atom uses only three of the four bonds that it is able to make. These electrons are called delocalized electrons, and they are free to move about in the layer. Graphite conducts electricity because these moving electrons can carry electric charge through the graphite. It is very unusual for a nonmetal to conduct electricity.

Mining and manufacture

Sticks of lead metal were used to draw on paper until 1564, when a deposit of very pure graphite was found in Borrowdale, England. The graphite was mined, cut into shape, and used for pencils. It was so valuable that the mine was open only a few weeks a year, and armed guards followed the wagons carrying the graphite! At the time, nobody knew that the substance was a form of carbon, so it was called plumbago, meaning "acts like lead." Over two hundred years later, Carl Scheele discovered that plumbago was made of carbon, not lead. Abraham Werner suggested the name graphite in 1789, from the Greek word that means "to write." People still talk about pencil lead, even though it is actually a form of carbon.

Graphite is found in metamorphic rocks such as marble and schist, and world production is about 600,000 tons a year. About 41 percent of the graphite comes from China, and much of the remainder comes from India, Brazil, and Mexico.

The uses of graphite

Graphite has a very high melting point, so it is used to make linings for furnaces, brakes for cars, and equipment to cast molten steel. It is also used as a heat-resistant lubricant because it is slippery. Since it conducts electricity, graphite is used in batteries, motors, and to make the electrodes needed to extract aluminum from its **ore** using electricity. Composite materials containing stiff graphite are used to make carbon-fiber boats, car parts, and sports equipment such as fishing rods and golf clubs.

◀ *This piece of graphite is from a plumbago mine. Graphite is a soft substance that flakes easily and feels greasy.*

Fullerenes

Three scientists discovered a new **allotrope** of carbon in 1985. Richard Smalley and Robert Curl from the United States, and Harold Kroto from Britain, fired a powerful laser at graphite and then analyzed the pieces that were blasted off. They detected a molecule made of 60 carbon **atoms,** C_{60}. When they worked out the arrangement of the atoms in this molecule, they found that it was a hollow sphere made of pentagons and hexagons—just like a soccer ball! They called the new molecule buckminsterfullerene, after an American named Richard Buckminster Fuller who designed geodesic domes (buildings that happen to look similar to the molecule). The different types of these molecules that have been discovered are called fullerenes for short. The three scientists won the 1996 Nobel Prize in Chemistry for their discovery.

▲

This is one of Buckminster Fuller's geodesic domes. These domes enclose large volumes using fewer building materials than normal buildings.

Buckyballs

Fullerenes are now made by passing electricity between a pointed graphite rod and a graphite disk. As the electricity jumps between them, it flashes like lightning and makes fine soot. When the soot is dissolved in a liquid called toluene, it makes a red-brown solution of fullerenes. If the solution is filtered and the toluene evaporated, almost pure fullerenes are left behind. The smallest fullerene molecule has 20 carbon atoms, and the largest so far has 540 carbon atoms. They are often called buckyballs because they are made from rounded cages of carbon atoms.

Research into buckyballs has led to new materials such as diamond films, and they have even been joined together to make **polymers.** With their rounded shape, buckyballs should act like tiny ball bearings, making them good lubricants.

The space inside buckyballs can contain other substances, such as metal atoms. These buckyballs, called fullerides, can even be shrunk to fit snugly around the metal atom inside!

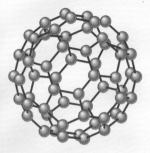

This model of a buckyball contains 60 carbon atoms. Its faces are hexagons and pentagons.

One or more of the carbon atoms in the buckyball can be traded for another **element** to make a huge range of molecules called fulleroids. Scientists are hopeful that these molecules will lead to new **catalysts** and electronic components. Fulleroids containing potassium or rubidium conduct electricity. In fact, they do it so well that at very low temperatures they are superconductors. This means that no energy is lost as heat when they conduct electricity, unlike normal conductors such as copper.

Nanotubes

If buckyballs are opened up, fullerene cylinders called nanotubes are made. These are very tiny tubes made of carbon that are thousands of times finer than a human hair. Nanotubes promise to be even more useful than buckyballs because they are six times lighter than steel, but a hundred times stronger. This means that they can be used in place of graphite fibers to make even lighter and stronger materials. It is already possible to buy a nanotube-reinforced tennis racket. Superconducting nanotubes have been developed, and theoretically they should eventually lead to wires that carry electricity without losing any energy.

▼ *This diagram shows part of a nanotube. One million nanotubes might be only 0.04 in. (1 mm) high.*

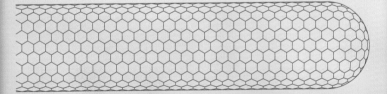

Coal

Coal is a sedimentary rock made from the ancient remains of plants. It is a natural source of **amorphous** carbon. Coal is commonly used as a fuel, especially in power stations, but it is also used to make a wide range of chemicals.

The formation of coal

During the Carboniferous Period, 280 million to 300 million years ago, huge swampy tropical forests covered a lot of the world. When the trees and other plants died, many of them did not rot away as usual because the swampy conditions stopped fungi and microorganisms from feeding on them. Thick deposits of dead plants built up on top of each other and were buried by layers of mud and sand. Over thousands of years, the weight of this mud and sand squashed the buried plants, and chemical **reactions** turned them into layers, or seams, of coal. Each foot (0.3 meter) of coal came from about 15 feet (4.5 meters) of dead plant material.

The equation for one of the reactions that turns cellulose in plant cell walls into carbon is:

$$\text{cellulose} \rightarrow \text{carbon} + \text{carbon dioxide} + \text{methane} + \text{water}$$
$$(C_6H_{10}O_5)_n(s) \rightarrow 4C(s) + CO_2(g) + CH_4(g) + 3H_2O(l)$$

The carbon dioxide, methane, and water gradually escape to leave the carbon behind as coal.

There are different types of coal depending on how long these reactions have been going on. Low-rank coals, such as dark brown lignite, contain the least amount of carbon and have formed relatively recently. High-rank coals, such as shiny black bituminous coal, contain the most carbon. Anthracite is the oldest and best coal of all—over 95 percent of it is carbon.

The United States and China mine over half of the total of mined black coal. Brown coal is mined mainly by Germany and Greece.

Coal mining

While the coal formed, the mud and sand that covered it turned into sedimentary rocks such as sandstone and shale. The coal seams are found under these rocks. Different methods are used to mine the coal depending upon how deeply it is buried.

Coal seams within 260 feet (80 meters) of the surface are usually removed by open-cast mining. The soil and rock on top of the coal is removed and stored. The coal is dug up using huge machines. Deep pits or strips are dug, depending on the layout of the coal seam and obstacles such as villages. After the coal has been removed, the stored rock and soil is put back. New plants are planted to reduce the damage done to the environment.

If the coal seams are too deep for open-cast mining, shafts are dug down to the coal. Miners dig into the seam—sometimes only 3–6 feet (1–2 meters) deep—and bring the coal to the surface through the mine shaft. Originally, miners used pickaxes and brute strength to remove the coal, and pit ponies pulled the coal wagons. It was dangerous work, but even children were sent down to work in the coal mines. Modern mines use machines to cut the coal and conveyor belts to transport it.

◀ Anthracite coal, shown here, is formed from fossilized plants from the Carboniferous Period (about 300 million years ago).

Uses of Coal

Coal as a fuel

Only 5 percent of the coal produced is used in homes for cooking and heating. The biggest use of coal is for generating electricity, and coal-fired power stations use 62 percent of the world's coal production. Crushed coal is burned in boilers in the power station, producing heat that is used to boil water to make steam. The steam drives a turbine that turns a generator to make electricity. These power stations produce a lot of electricity. One of the generating units at Paradise Power Station in Kentucky produces over 1,000 megawatts, or 10 billion kilowatt-hours per year. Drax power station in North Yorkshire in the United Kingdom (the biggest in Europe) has six generating units that produce a total of 4,000 megawatts—enough to run 40 million 100-watt light bulbs continuously!

Although coal-fired power stations produce 37 percent of the world's electricity, they are not very efficient. Only 35 percent of the energy in the coal is converted into electricity—the rest escapes as heat into the environment. When coal burns, the carbon in it reacts with oxygen in the air to make carbon dioxide. This is a greenhouse gas that traps heat in the atmosphere, causing global warming.

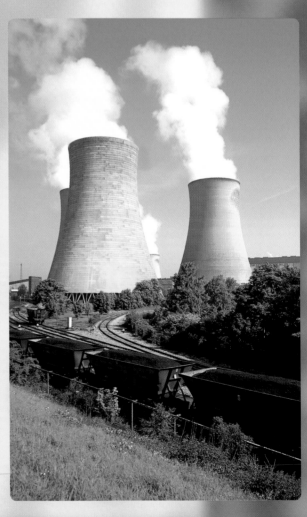

Train cars bring coal to be burned in this coal-fired power station. ▶

Coal contains sulfur impurities that produce sulfur dioxide when the coal is burned. If this gas escapes into the atmosphere, it dissolves in the water in the clouds to produce acid rain. Coal also contains various **minerals** that leave ash behind. Some ash escapes out of chimneys. However, in modern coal-fired power stations, filters remove most of the sulfur dioxide and ash before they can escape into the atmosphere.

Coal in the chemical industry

If coal is heated very strongly while keeping the air away, it breaks down to make some useful new substances. This process is called **destructive distillation,** and it produces coal gas, coal tar, and coke.

Coal gas, sometimes called town gas, is a smelly mixture of gases including methane, carbon monoxide, and hydrogen. Town gas was used for heating and lighting before natural gas was discovered. It is still used today as an industrial fuel, and it can be used in metal **refining.**

Coal tar is a mixture of over a hundred different carbon **compounds,** such as benzene and naphthalene. These compounds are separated using **fractional distillation,** and are used to make Nylon®, varnish, dyes and paints, explosives, medicines, and pesticides. A thick black tar is left over after fractional distillation. This is used to waterproof roofs and to make roads.

Coke is pure carbon and makes a very good fuel. When it burns, it releases more heat than coal and much less smoke. It is also used as a starting material in the chemical industry for making plastics and other substances. Coke is one of the raw materials used in the extraction of iron from iron **ore** in blast furnaces.

Carbon Dioxide

Carbon dioxide is a clear, colorless gas that is denser than air. There are 2,800 billion tons of carbon dioxide in the atmosphere. It dissolves easily in water, so the oceans contain about fifty times more carbon dioxide than the atmosphere does.

Combustion

Combustion is the chemical word for burning—the chemical **reaction** that occurs when fuels react with oxygen in the air. Fuels that contain carbon include coal, wood, natural gas, and fuels made from crude oil, such as gasoline and diesel. They all release carbon dioxide when they burn. Coal is made almost entirely of carbon, so it produces the most carbon dioxide when it is burned. The other fuels also contain hydrogen and other **elements.**

The equations for combustion are:

1) coal + oxygen → carbon dioxide
$$C(s) + O_2(g) \rightarrow CO_2(g)$$
2) fuel + oxygen → carbon dioxide + water

The second equation works for fuels that also contain hydrogen, such as wood, natural gas, and gasoline.

Carbon dioxide stops fuels from burning and is used in fire extinguishers. These fire extinguishers are safe to use on electrical fires because they do not use water.

If the supply of oxygen is limited, incomplete combustion happens. Some of the carbon in the fuel is released, making the flame smoky and sooty, and some of it reacts with oxygen to produce carbon monoxide instead of carbon dioxide. Carbon monoxide is a poisonous, colorless gas with no smell. Small amounts of it will cause you to fall asleep, and larger amounts can kill you.

◄ *Limewater is an alkaline liquid used to detect carbon dioxide. In this experiment, the limewater has turned cloudy white, showing that the boy breathed out carbon dioxide.*

Respiration

Respiration is the chemical reaction that every cell in our body uses to release energy from food. Without it, we could not get the energy we need for all our body's processes, such as moving, growing, and keeping warm.

The equation for respiration is:

glucose + oxygen → carbon dioxide + water
$$C_6H_{12}O_6(aq) + 6O_2(g) \rightarrow 6CO_2(g) + 6H_2O(l)$$

This happens in tiny objects in our cells called mitochondria.

Blood carries waste carbon dioxide to our lungs, where it is expelled when we breathe out. The air we breathe in contains about 0.036 percent carbon dioxide, but the air we breathe out contains about 3.7 percent carbon dioxide when we are resting, to more than 5 percent when we are exercising.

Fermentation

Yeast are microscopic single-celled fungi. They contain **enzymes** that can break up sugar to release energy in a process called **fermentation.**

The equation for fermentation is:

$$\text{glucose} \xrightarrow{\text{enzymes in yeast}} \text{ethanol + carbon dioxide}$$
$$C_6H_{12}O_6(aq) \longrightarrow 2C_2H_5OH(l) + 2CO_2(g)$$

Yeast is used in making bread because the carbon dioxide released makes the dough rise. Yeast is also used in making wine and beer because fermentation releases ethanol (alcohol).

More About Carbon Dioxide

Carbon dioxide is added to drinks under pressure to make them bubbly. When drinks like champagne and cola are opened, the pressure is released and lots of carbon dioxide bubbles escape.

If carbon dioxide is cooled to −109.3 °F (−78.5 °C), it freezes to make a white solid called "dry ice." This is used to keep food and medical samples cold while they are transported. As it warms up, the dry ice sublimes, or turns back into a gas without becoming a liquid. It creates a white mist of condensed water vapor that is used for special effects in rock concerts and films.

Dry ice dropped into a beaker of water ▲ makes bubbles and lots of mist that flows over the edge.

Photosynthesis

Photosynthesis is the chemical **reaction** in plants that produces food by using carbon dioxide, water, and energy from sunlight.

The equation for photosynthesis is:

$$\text{carbon dioxide + water} \xrightarrow{\text{sunlight}} \text{glucose + oxygen}$$

$$6CO_2(g) + 6H_2O(l) \longrightarrow C_6H_{12}O_6(aq) + 6O_2(g)$$

Photosynthesis is carried out in the plant cells in tiny green objects called chloroplasts.

Plants also need to respire. During the day, when there is a lot of sunlight, the rate of photosynthesis is more than the rate of respiration. It is then that plants take in carbon dioxide. However, at night when it is dark, plants cannot photosynthesize, so they release more carbon dioxide than they take in as they respire. Farmers often add extra carbon dioxide to the air if they are growing their crop in greenhouses. This makes the plants photosynthesize faster, so they grow faster and produce more food.

The greenhouse effect

Energy from the Sun passes through the atmosphere, and some of it escapes back into space as infrared radiation, or heat. The carbon dioxide in the atmosphere is very good at absorbing infrared radiation, trapping heat in the atmosphere. This is called the greenhouse effect, and it keeps Earth warm. The average surface temperature of our planet is 57.2 °F (14 °C), but without the greenhouse effect Earth would be almost as cold as the Moon, with an average surface temperature of −0.4 °F (−18 °C).

Human activities have been adding more carbon dioxide to the atmosphere than can be removed by natural processes. There are two main activities involved. Industrial processes and burning **fossil fuels** produce 65 percent of the extra carbon dioxide, and cutting down forests for building and farmland produces most of the rest. Fewer trees also mean that less carbon dioxide can be removed by photosynthesis. In 300 years, the amount of carbon dioxide in the air has gone up from 0.0275 percent to 0.0365 percent. This might not sound like a lot, but as the levels have gone up, so has the temperature at Earth's surface. Average temperatures have increased about 1.08 °F (0.6 °C) since the late 19th century. This is called global warming, and it is leading to changing weather patterns. The polar ice caps may melt, causing sea levels to rise and lowland areas to flood.

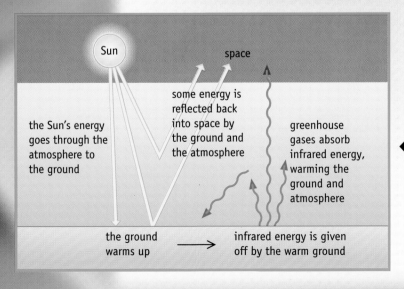

Sun

space

the Sun's energy goes through the atmosphere to the ground

some energy is reflected back into space by the ground and the atmosphere

greenhouse gases absorb infrared energy, warming the ground and atmosphere

the ground warms up

infrared energy is given off by the warm ground

This diagram shows how carbon dioxide traps heat in the atmosphere. Methane from rice fields and flooded land is another greenhouse gas.

Calcium Carbonate

Calcium carbonate is a solid **compound** that contains calcium, carbon, and oxygen. Living things in the oceans, such as snails and crabs, can use carbon dioxide dissolved in the water to make calcium carbonate for their shells. Vast numbers of microscopic animals and plants called plankton also make their shells from calcium carbonate. When these living things die, they sink to the seabed to form layers of shells. These layers, called sediments, build up on top of each other, and their weight squeezes the water out from between the shells. Over thousands of years, the shells become stuck together by salt crystals to form sedimentary rocks called limestone and chalk.

Chalk

Chalk is a white, crumbly rock. Gymnasts and weightlifters rub it onto their hands to get a better grip. Cement is made by roasting a mixture of chalk and clay. Concrete is made by mixing cement, sand, aggregate (small stones), and water together.

Limestone

Limestone is a tough rock that can be used as a building material, and crushed limestone is used in road building. Carbonates, such as the calcium carbonate in limestone, **neutralize** acids by reacting with them to produce a calcium salt, water, and carbon dioxide. This can be useful in areas where there is a risk of high amounts of acid rain. Acid rain can pollute lakes, making it difficult for fish and other living things to survive. Adding powdered limestone to the lakes can help neutralize the extra acid.

If limestone is heated, it decomposes (breaks down) to make calcium oxide and carbon dioxide. Calcium oxide is often called lime, and farmers use it to neutralize acid soils to help their crops grow better. Mortar, used by builders to join bricks together to make walls, is made by mixing lime with sand and water. When iron is being made from iron **ore** in the blast furnace, it contains sandy impurities. Calcium oxide from heated limestone reacts with these impurities to make a slag that is easily removed, leaving pure iron.

Marble

If limestone is buried deep underground by Earth movements, it gets squashed and heated, and turns into a metamorphic rock called marble. This is shiny and harder than limestone or chalk. It can be polished to an attractive finish and is used for statues, for kitchen and bathroom counters, and on the outside of buildings as decoration.

Acid attacks rock

Calcium carbonate reacts with acid and dissolves away. Rain is naturally acidic because carbon dioxide dissolves in it to make carbonic acid. Over thousands of years, the acid in rain and rivers dissolves limestone and chalk to form caves. When water drips from the roof of a cave, it evaporates to leave columns of calcium carbonate behind. These are called stalactites and stalagmites. Air pollution has made rain more acidic than normal, causing it to badly damage statues and buildings made from limestone or marble.

◀ *Acid in rainwater has reacted with limestone to form this cave. The columns of calcium carbonate hanging from the ceiling are called stalactites, and the ones extending upward are called stalagmites.*

Oil and Natural Gas

Oil and natural gas contain **compounds** of hydrogen and carbon called **hydrocarbons.** They are used widely as fuels, but they are also used to make **products** such as plastics, dyes, and cosmetics. The oil and atural gas on Earth will eventually run out, so these re called nonrenewable energy resources.

The formation of oil and natural gas

Millions of years ago, sea creatures died, sank to the seabed, and were covered by layers of **mineral** sediments. The sediments and the remains of the dead creatures were buried deeper and deeper. Oxygen could not get to the remains, so they did not rot away. Instead, they heated up and were squashed by the weight of the sediments. Eventually, the mineral sediments turned into rock, and the remains turned into oil and gas. Water pushed the oil and gas upward through **permeable rock** (rock with holes and cracks like a sponge). When they reached **impermeable rock,** they could not rise any further and were trapped.

▲ *Crude oil is not at all like the engine oil used in cars. It is thick, black, and very smelly.*

Drilling for oil

Oil workers drill deep into the ground and through the layer of impermeable rock to reach oil and gas. Natural gas is usually found just below the impermeable rock, while oil is found further down. Once it is out of the ground, a pipeline or tanker sends the oil to an oil refinery.

Oil refining

At the oil refinery, oil is separated into different hydrocarbons by a process called **fractional distillation.** The oil is heated to about 662 °F (350 °C) and pumped into the bottom of a tall metal tower called a fractionating column. The column is very hot at the bottom, and vapors from the heated oil rise up the column towards the top where it is colder.

Hydrocarbons with many carbon **atoms** have very high boiling points. They remain solid at 662 °F (350 °C), so they just stay at the bottom of the column. Hydrocarbons with fewer than five carbon atoms in them are gases. They have low boiling points, so they are able to make it right to the top without condensing into liquid. The vapors from the medium-sized hydrocarbons rise until they get cool enough to condense back into liquids. The liquids are collected in trays inside the column and piped away.

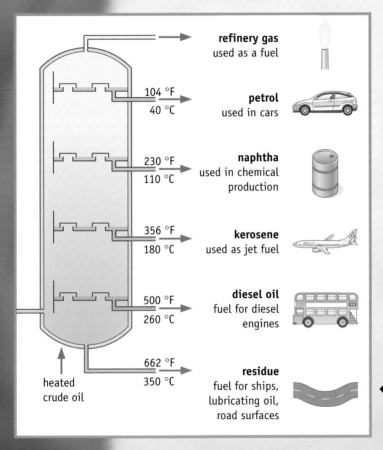

refinery gas
used as a fuel

104 °F
40 °C
petrol
used in cars

230 °F
110 °C
naphtha
used in chemical production

356 °F
180 °C
kerosene
used as jet fuel

500 °F
260 °C
diesel oil
fuel for diesel engines

662 °F
350 °C
residue
fuel for ships, lubricating oil, road surfaces

heated crude oil

The different parts separated from the oil are called fractions. The fractions at the bottom of the column are solids with big molecules, those in the middle are liquids with medium-sized molecules, and the fractions at the top are gases with very small molecules.

◀ *These are the main fractions derived from oil.*

Cracking

Crude oil often contains too many big molecules but not enough of the more useful medium-sized molecules, like gasoline. The big molecules are converted into smaller molecules by using a **catalyst** or by heating them strongly under pressure. This process, called cracking, is important because some of the smaller molecules are useful to the chemical industry and can be used to make plastics.

Complex Molecules

Carbon **atoms** can form four **bonds** with other atoms, including carbon itself. This means that carbon atoms can join together to make chains and rings of carbon atoms, with other **elements** joined on in countless different ways. Living things produce large and complex molecules by joining small molecules together. Chemists have learned how to copy this to make synthetic materials such as plastics and medicines.

Plastics

When oil fractions are cracked in the refinery, **alkenes** are made. Alkenes contain two or more carbon atoms joined by double covalent bonds, and they can join end to end to make very long molecules called addition **polymers.** Lots of addition polymers are possible. For example, ethene molecules can join end to end to make polyethylene, used to make bags; propene molecules make polypropylene, used to make tough ropes; and styrene molecules make polystyrene, used for fast-food containers and packing materials.

▼ *All these products are from polymers—long, complex carbon compounds made from lots of small molecules called monomers.*

Carbohydrates

Carbohydrates such as sugar and starch are **compounds** of carbon, hydrogen, and oxygen that occur naturally. Two molecules of a simple sugar can join together to make more complex sugars. Sucrose (cane sugar) is made from glucose and fructose joined together, and lactose (found in milk) is made from glucose and galactose. Thousands of glucose molecules can join together to make starch (found in rice and bread) or cellulose, depending on how they are joined. Cellulose is the tough molecule found in plant cell walls.

Polyesters

Fats and oils are made from two carbon compounds, glycerol (a type of alcohol) and fatty acids. The fatty acids join on to the glycerol using a type of chemical bond called an ester bond. Some polymers chemists make are called polyesters because they contain lots of ester bonds. Polyester can be used to make bottles and videotape. Polyester fibers, such as Terylene®, are used to make soft, durable clothes.

DNA

DNA stands for deoxyribonucleic acid—a long name for often a very long molecule. A single DNA molecule can be 2.75 inches (7 centimeters) long, and each human cell contains 6.6 feet (2 meters) of it tightly coiled to fit inside the **nucleus!** DNA contains the genetic code for producing all the different proteins a cell needs, and human DNA contains at least 30,000 genes. DNA is made from four different carbon compounds called nucleotides, that can join together in countless combinations.

Proteins and polyamides

Proteins are large molecules made from small carbon compounds called amino acids. Amino acids join to each other using a type of chemical bond called a peptide bond. Polyamides contain the same sort of bond. Polyamides, such as Nylon®, are used to make ropes, shirts, and ladies' tights.

The Carbon Cycle

How many carbon **atoms** are there in your body? If you have a mass of 132 lb (60 kg), your body will contain around 31 lb (14 kg) of carbon. That's about 690 million billion billion carbon atoms, and there's a chance that one of those atoms could once have been part of a scientist like Mendeleev, or of a prehistoric plant! This is because different processes continually recycle carbon between the atmosphere and the other substances that contain carbon—including you!

Processes that remove carbon dioxide from the atmosphere

In the process of **photosynthesis,** green plants take in carbon dioxide from the atmosphere and make glucose from it. This is used to make starch, cellulose, proteins, fats, and all the other complex molecules a plant needs to stay alive. Carbon dioxide from the atmosphere dissolves easily in water to form carbonic acid and metal carbonates. Some of this is used by sea creatures to build their shells, which eventually become sedimentary rocks such as chalk and limestone. This is another example where carbon dioxide from the atmosphere enters into the processes of living things.

Processes that return carbon dioxide to the atmosphere

Plants and animals respire as they use energy from their food, and this returns carbon dioxide to the atmosphere. It's a strange thought, but the energy we use to move or grow was originally sunlight, captured by the green chlorophyll in plant leaves. When animals and plants die, other organisms feed on them and they also respire. If dead things did not decay, we would quickly be knee-deep in them! Some dead animals and plants form **fossil fuels** instead of decaying, and these release carbon dioxide when they are burned. The energy released when we burn coal, oil, or natural gas was originally sunlight, captured by plants millions of years ago. Carbon dioxide is also released when limestone is used in the manufacture of cement and iron.

Global warming

Gases such as carbon dioxide and methane trap heat in the atmosphere, causing Earth to warm up. The carbon cycle should be balanced, but it is not. Human activities are releasing more carbon dioxide than can be used up naturally, causing global warming. Large amounts of carbon dioxide are released when fossil fuels are burned, but clearing and burning forests for farmland adds to the problem. This is because the burning trees release carbon dioxide, and there are fewer trees left for photosynthesis.

▼ *This is the carbon cycle. It shows how carbon in all sorts of **compounds** is recycled.*

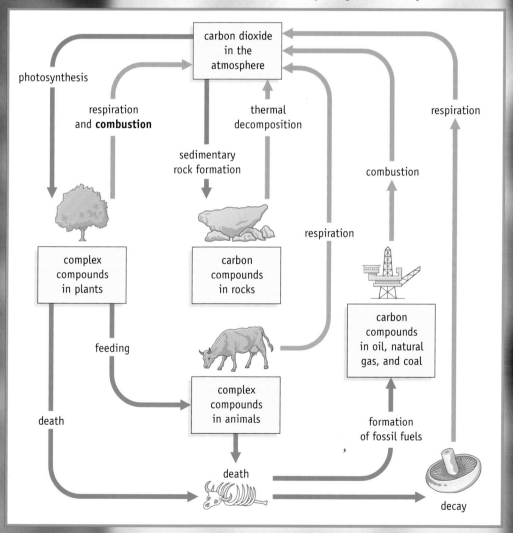

CHAPTER 4

Silicon

Silicon makes up 25.7 percent of Earth's crust and is the second most abundant **element** in it. However, it was not discovered until 1823 because it is not found as a free element. Instead, silicon is usually found joined with oxygen to make silicon dioxide, or silica. Silica can be used to make quartz, and it is the most common **mineral** in sand. Silicon is also found joined with oxygen and other elements to make silicates. There are more than 1,000 silicates, including talc, clay, and asbestos. Silicon is widely used in electronic devices, and silicon **compounds** are used in concrete, brick, pottery, glass, and silicone sealants. Silicon carbide (carborundum) is the next hardest substance after diamond and is used to make cutting tools.

The structure of silicon

There are two forms, or **allotropes,** of silicon. **Amorphous** silicon is a red-brown powder that does not conduct electricity. Crystalline silicon is a shiny gray-black solid that conducts electricity, though not as well as a metal does. The structure of crystalline silicon is very similar to the structure of diamond. A crystal of crystalline silicon is a single giant molecule, with every silicon **atom** joined to four other silicon atoms using covalent **bonds.** Because each bond is strong and there are very many of them, silicon is hard and has a high melting point, 2,570 °F (1,410 °C).

Silicon is fairly unreactive. It does not react with water or most acids, but it will react with steam and alkalis. It will also react with oxygen to make silicon dioxide (silica). This exists in two forms: crystalline silica, such as quartz, and amorphous silica, such as opal.

The manufacture of silicon

There are three stages in the manufacture of silicon. Sand (silicon dioxide) is the usual starting material, and the end result is a fairly pure form of the element. However, this is not pure enough for the **semiconductor** materials used to make silicon chips. A process called **zone refining** is used to make hyperpure (very pure) silicon for electronic circuits.

▲ *These huge sand dunes contain countless grains of sand, a compound of silicon and oxygen.*

In the first stage of extracting silicon, the oxygen is removed from the sand by heating it with coke (a cheap type of pure carbon) to leave impure silicon behind. The equation for the first stage in making silicon is:

silicon dioxide + carbon → silicon + carbon dioxide
$$SiO_2(s) + C(s) \rightarrow Si(s) + CO_2(g)$$

The silicon made at this stage is impure. In the second stage, the impure silicon is reacted with chlorine gas to make a liquid called silicon tetrachloride. This is purified using **fractional distillation.** The equation for the second stage in making silicon is:

silicon + chlorine → silicon tetrachloride
$$Si(s) + 2Cl_2(g) \rightarrow SiCl_4(l)$$

In the third stage, the pure silicon tetrachloride is reacted with hydrogen to make silicon. The equation for the third stage in making silicon is:

silicon tetrachloride + hydrogen → silicon + hydrogen chloride
$$SiCl_4(l) + 2H_2(g) \rightarrow Si(s) + 4HCl(g)$$

Chips

Computer chips and other electronic devices are made from **semiconductors.** These are substances, such as silicon, that are electrical insulators at room temperature, but conductors when they are warmed up. They also conduct when tiny amounts of other **elements** are added to them, a process called doping. If silicon is doped with phosphorus or arsenic, it makes N-type silicon. Electrical charge moves through N-type silicon using **electrons,** which are negatively charged. If silicon is doped with boron or gallium, it makes P-type silicon, and electrical charge moves through it using positively charged "holes." The millions of transistors in computer chips are made from both types of doped silicon.

Wafers in a "clean room"

Computer chips start as very pure, or hyperpure, silicon. Single crystals of hyperpure silicon are made into rods up to 12 in. (30 cm) in diameter, then sliced into circular wafers

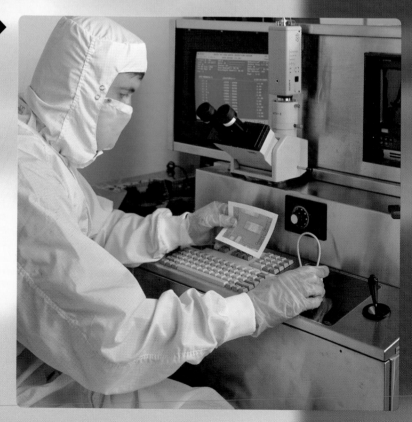

A technician seated at a microscope in a "clean room" holds a photograph of part of a silicon wafer while checking on its microcircuit.

only 0.02 in. (0.5 mm) thick using a diamond saw. The wafers are polished flat and checked for scratches. It is very important that dust does not get onto the wafers while the electronic components are being made on them. Even one speck of dust could make a computer chip useless, For this reason, the chips are made in a "clean room."

A cubic meter (1.3 cubic yards) of ordinary air contains millions of dust particles, but the air in a "clean room" has fewer than forty specks of dust in that same amount of air. To stop dust, skin, and hair from getting into the room, everyone working in there must wear special protective clothing called a **bunny suit** over his or her normal clothes. This suit includes a helmet with a battery-powered air filter, a hair net, two layers of nylon and latex gloves, and two layers of overshoes. Even beards have to be covered up!

Chip building

Lots of chips are made on a single wafer. A typical chip might contain 20 layers, each with a complex pattern of lines and shapes etched in silicon. Chip designers make a set of special stencils called masks, and each mask has the pattern for one layer. Ultraviolet light is used to transfer the pattern from a mask onto the wafer in a process called **photolithography.** Different chemicals are used to shape the layer and to dope the silicon. The process is repeated with different masks to build up the layers, one on top of the other.

Copper and aluminum are added to conduct electricity from one part of the chip to another. Each chip is tested while it is still on the wafer. Then the wafer is carefully cut up into separate chips using a diamond saw. Each chip is put into a plastic case to protect it, and very fine gold wires are used to connect it to circuit boards.

Super Sand

Sand is largely made up of crystalline silicon dioxide. It is used in concrete and mortar, and to make sandpaper. Dirty buildings can be cleaned by blasting sharp sand through a hose using compressed air. Finely powdered sand is used in paints, plastics, and **ceramics.** Sand mixed with clay is used to make bricks, as well as molds for casting metal. Pure sand is used to line furnaces because it is heat-resistant, but the biggest single use for sand is in making glass. About 38 percent of the 107,000 tons of sand used in the world each year is used in this way.

Making glass

If sand is melted and then cooled slowly, it turns back into solid crystals of sand. However, if the liquid sand is cooled very quickly, it solidifies before the crystals can form, and glass is made instead. Volcanic eruptions often provide the conditions needed for glass to form naturally, and prehistoric people used glass made this way to make tools. The first man-made glass was created about 6,000 years ago.

For nearly 2,000 years, mouth blowing was the main method of forming glass objects, until machines took over in the last century. The glass blower shapes the glass by turning the pipe continuously, blowing air into the glass, and using a wet graphite mold.

Glass can be made from sand alone. However, there are two problems: a very high temperature is needed, and the resulting glass melts in water. If sodium carbonate is added to the sand, the melting temperature is reduced from 3,092 °F (1,700 °C) to about 1,472 °F (800 °C). If calcium carbonate and magnesium carbonate are added to the glass while it is molten, this prevents it from dissolving. Glass made in this way is called soda-lime glass or bottle glass. Unfortunately, it shatters easily if hot liquids touch it.

Borosilicate glass does not shatter easily when heated because it contains other chemicals, such as boric acid. This kind of glass is used in laboratories, fiberglass, and heat-resistant cooking bowls. Another type, aluminosilicate glass, contains about 20 percent aluminum oxide. It is more heat-resistant than borosilicate glass, so it is used in halogen lamps, that become very hot indeed. Glass ceramics are particularly heat-resistant, so they are used for stove burners and missile nose cones.

Lead crystal glass contains at least 24 percent lead oxide. It can be cut and engraved easily to make attractive glassware. Although lead **compounds** are poisonous, they are trapped in the glass, so there is usually little chance of being poisoned when drinking from it. Glass with a lot of lead oxide (over 65 percent) is used for radiation shielding.

Silicosis

Tiny particles of silica from sand can cause a lung disease called silicosis. The lungs of someone who breathes in silica dust over a long time develop scars and tough bumps called nodules. If the nodules get too big, it becomes difficult to breathe, and the victim may die. To avoid these dangers, workers have to take precautions to prevent breathing in any silica. This is serious in jobs such as sandblasting, drilling and blasting rock or concrete, demolition of buildings, and mining.

Diatomite, Silicates, and Silicones

Diatomite

Diatoms are tiny single-celled plants that live in rivers, lakes, and seas. They are so small that 25 million of them could fit into a teaspoon! They are extremely useful to us because they make their shells from silica. When they die, their shells sink and form a sedimentary rock called diatomite. This rock is largely made up of **amorphous** silica, along with some other chemicals. About 2 million tons of diatomite are used every year, mostly for filtering beer, oil, medicines, and swimming pool water. It is also used in paints, plastic, rubber, and paper, as well as for making silicates. Diatomite is very good at absorbing liquids, so it is used for all sorts of purposes, from soaking up chemical spills to making cat litter!

Silicates

Silicon dioxide will not dissolve in water—try dissolving some sand in a beaker of water as an experiment. However, it **reacts** with sodium hydroxide (a common alkali) to make sodium silicate, a substance that does dissolve in water. Sodium silicate is a solid containing sodium, silicon, and oxygen. It is often called "water glass," and you may have seen it if you have made a "crystal garden" at home or school.

There are other silicates that do not dissolve in water. Calcium silicate is used for heat insulation and sound insulation. Magnesium silicate is a soft and slippery solid, often called French chalk or talc. It is the main ingredient in talcum powder, and it is also used in soap, paint, and fireproof materials. If you have ever had a tooth filled by the dentist, zirconium silicate might be an ingredient in your filling.

Zeolites are porous solids made from aluminosilicates (these are silicates that contain aluminum). Natural zeolites are mainly used in cat litter, but they can also be used to turn hard water into soft water by a process called ion exchange. Hard water contains calcium or magnesium salts that make it difficult to get soap to lather. Zeolites soften the water by removing these salts. Man-made zeolites are used as **catalysts** in oil refineries to crack oil fractions.

Slippery silicones

Silicones were discovered in 1904 by an English chemist, Frederick Kipping. These are **compounds** that contain long chains of silicon and oxygen **atoms,** with carbon and hydrogen atoms attached to them. Silicones are colorless, unreactive substances that repel water (they don't get wet). This means that they are very useful for waterproofing raincoats and for sealing the edges around bathtubs and showers. Some silicones are oily liquids used in cosmetics, while others are used to lubricate moving parts and to conduct heat away from those parts in machinery. Some silicones do not squash easily, so they are used as hydraulic fluid in car braking systems. Silicones can also be made into the rubbery solids used in cosmetic surgery.

◀ *This "crystal garden" sits in a solution of "water glass." Crystals of metal salts such as copper sulfate are dropped into the "water glass" and left for about a week. Colored hollow tubes made from metal silicates grow and branch from the original crystals.*

Germanium

During the 18th and 19th centuries, chemists were particularly busy discovering new **elements.** To make things easier for themselves, chemists desperately needed a way to put all the elements into some sort of order. It was a bit like attempting to do a difficult jigsaw puzzle without a picture to help you!

In 1829, a German chemist named Johann Döbereiner noticed that some elements shared similar properties. He was able to put these elements into groups of three, called triads. However, chemists kept finding new elements that did not fit his triad pattern.

By 1863, the number of elements identified totaled 56. At this point, an English chemist named John Newlands began to organize the elements. He put them in order of their atomic weight (how heavy their **atoms** were compared to each other), and then put them into rows of seven elements. Newlands found that for many elements, the eighth element that followed it behaved in a similar way. He called his discovery the Law of Octaves. Unfortunately, his law did not work for all the elements, and many scientists just mocked his ideas. Throughout this time, chemists kept finding even more elements, and all of them needed to be ordered.

▶ *Dimitri Mendeleev spent much of his life in St. Petersburg, Russia, where you can see his statue and a wall showing his periodic table.*

Six years later, Dimitri Mendeleev published his first **periodic table**, later updated in 1871. Mendeleev arranged the elements in order of their atomic weight, just as Newlands had done, but he did two important things that made his table work. He realized that since so many new elements had been discovered, there must be many others waiting to be found. Mendeleev left gaps for undiscovered elements in his periodic table. From their positions in his table he also predicted what they should be like. One of the gaps he left was between silicon and tin. Mendeleev called the missing element ekasilicon, meaning "below silicon." He predicted that ekasilicon's properties would be between those of silicon and tin.

In 1886, a German chemist named Clemens Winkler discovered a new element in an **ore** called argyrodite. He called this germanium—it was the real ekasilicon. The properties of germanium were very close to the properties that Mendeleev had predicted fifteen years earlier.

Property	ekasilicon	germanium
color	gray	gray–white
atomic weight	73.4	72.3
density of the element	0.199 lb/in.^{3} 5.5 g/cm	0.192 lb/in.^{3} 5.32 g/cm^3
density of the oxide	0.17 lb/in.^{3} 4.7 g/cm^3	0.17 lb/in.^{3} 4.7 g/cm^3
boiling point of the chloride	less than 212 °F less than 100 °C	186.8 °F 86 °C

▲ *This table shows some of Mendeleev's predicted properties for ekasilicon and the real properties of germanium. Germanium's melting point was 284.4 °F (158 °C) higher than Mendeleev predicted, but his other predictions were incredibly close.*

Support for new ideas in science comes from using good scientific reason to make a prediction that then comes true. This meant that the discovery of germanium was a great success for Mendeleev's periodic table. Our modern periodic table is based closely on his table.

Manufacture and Uses of Germanium

Germanium is a shiny gray-white solid that is hard, but easily shattered. Germanium is found combined with other **elements,** rather than on its own as a free element. Clemens Winkler discovered germanium in argyrodite (a compound of germanium, sulfur, and silver), but the main **ore** is germanite (a compound of germanium, sulfur, copper, and iron). Germanium is extracted from germanium (IV) oxide using a method similar to the one used to extract silicon. About 58 tons are produced every year.

Germanium and the discovery of the transistor

In the middle of the 1900s, vacuum tubes were used to make, control, and amplify electrical signals in electronic equipment. These were not very convenient because they were large, fragile, and took time to warm up before they could be used. Modern equipment uses transistors instead because they are much smaller and tougher, and do not need to warm up. Three American physicists—John Bardeen, Walter Brattain, and William Shockley—invented the first transistor in 1947. This was made from germanium, a **semiconductor** like silicon. The first integrated circuit (a sort of simple computer chip) was built onto a piece of germanium in 1958. This was difficult to do, and most modern chips use silicon instead. Germanium is still used in transistors and in high-powered electronics. However, this is no longer its biggest use—about half of the germanium produced is used for optical fibers.

Germanium and light

Optical fibers are very thin glass tubes, often just 0.005 inch (0.125 millimeter) thick. They consist of an inner core of glass surrounded by an outer layer of glass. Light shining into one end can travel through the core and will pass through to the other end with very little being lost on the way. The light is kept from escaping out of the sides by the outer layer. This process works because the glass in the core contains germanium. Germanium increases the "refractive index" of the glass and causes light to be reflected back into the core where the core and the outer layer are joined.

Endoscopes are tubes containing optical fibers. Endoscopes let doctors see inside their patients without surgery, and let engineers inspect the inside of machinery without taking it apart. Optical fibers are used in long-distance telephone lines and computer network cables. Signals are sent through them as pulses of infrared light in digital code. So little light is lost on the way that amplifiers (called repeaters) are often needed only every 62 miles (100 kilometers).

Crystals of bismuth germanate are used in gamma ray detectors because they glow when hit by radiation. Magnesium germanate is used in fluorescent lights because it glows when hit by ultraviolet light.

▲ Cables containing bundles of optical fibers like these carry large amounts of information, such as telephone conversations and computer data.

Tin

Ordinary tin is called white tin or β-tin (beta-tin). It is a soft, silvery-white metal with a bluish tinge. When it is bent, crystals in the metal break and you hear a noise called "tin cry." If white tin is cooled below 55.76 °F (13.2 °C), it gradually changes into another **allotrope** of tin called gray tin or α-tin (alpha tin). This form is powdery and not very useful. The change from white to gray tin is often called tin pest, because people used to think that the devil or microbes caused it!

Tin pest was one of the reasons why Napoleon's 1812 campaign against Moscow failed. Napoleon bought a million overcoats from England for his troops, and to save money he ordered tin buttons instead of brass ones. The buttons turned into gray tin and crumbled away in the cold Russian winter, making his troops more uncomfortable than they already were. Modern tin usually contains small amounts of bismuth or antimony to keep it from turning into gray tin.

These engine houses in the closed Botallack tin mines in Cornwall, England, were once used to pump out flooding water. Being close to sea level lessened the height needed for pumping, so old engine houses are often found perched on cliff tops. ▶

Cassiterite, or tin oxide, is the main ore of tin.

Tin smelting

Each year 200,000 tons of tin are produced in the world. Tin is not found in its native form as a free **element**, but it has been known for thousands of years because it is easy to extract from its **ore.** The main ore is cassiterite, or tin (IV) oxide. This is mined in many countries, although China, Indonesia, and Peru are the main producers. Extracting tin from its ore is called smelting.

To extract tin, crushed cassiterite is heated to about 1,112°F (600°C) to remove any trapped sulfur impurities. It is then heated to about 2,462°F (1,350°C) and stirred with coke for about 15 hours.

The equation for smelting tin is:

tin (IV) oxide + carbon → tin + carbon dioxide
$$SnO_2(s) + C(s) \rightarrow Sn(l) + CO_2(g)$$

Coke or anthracite coal is used to provide the carbon. Carbon is more reactive than tin, so it removes the oxygen from the tin (IV) oxide. This sort of reaction is called a displacement **reaction.**

A slag containing impurities eventually floats on a pool of molten tin, so it can be removed. The tin is poured into molds, cooled, and solidified. It is then purified or refined to remove impurities.

Refining tin

The most common method of **refining** tin is known as fire-refining. The tin is heated to about 2,192 °F (1,200 °C) in a vacuum. At this temperature, the tin melts and many of its impurities boil away, leaving tin that is up to 99.85 percent pure. If very pure tin is needed, electricity can be used to remove the impurities in a process called **electrolysis.** Tin refined this way can be 99.9999 percent pure!

Uses of Tin

Tin coatings and food cans

If tin is heated with steam or air, it reacts to make tin (IV) oxide, but it will not react with water or air at room temperature. This makes it very useful for coating other metals to keep them from corroding. Tin has a low melting point of 449.6 °F (232 °C), so metal objects can be coated just by dipping them in a bath of molten tin. Tin coatings stick very well and do not easily flake off, even when the object is bent or stretched.

Tinplate is steel coated with a layer of tin, often just 0.0004 in. (0.001 mm) thick, by a process called electroplating. The steel is dipped in a bath of tin (II) chloride or tin (II) sulfate solution, and electricity is passed through it. This causes a layer of tin to form on the surface of the steel. About 30 percent of the tin produced is used in tinplate, and nearly all of this goes to make food cans. Thin steel is used in modern cans, but the first cans were so thick that a hammer and chisel were needed to open them!

The Liberty Bell in Philadelphia was originally cast in Britain in the mid-1700s. Its metallic composition consists of 70 percent copper, 25 percent tin, and small amounts of lead, zinc, cassiterite, gold, and silver, with traces of antimony and nickel.

Alloys and alloy coatings

Tin is used in a large number of alloys, such as solder—an **alloy** of tin and lead. Solder has a low melting point and is used to make electrical connections. Pewter, another alloy, is used to make attractive tableware, such as drinking goblets and candlesticks. Traditional pewter contains lead, but modern pewter is mostly tin with small amounts of copper, bismuth, and antimony to harden the tin. Bronze is also a tin alloy. Different bronzes can be made by varying the basic mixture of tin and copper, or by adding other metals.

Metals can be coated with tin alloys for protection and decoration. Tin and nickel alloy coatings resist corrosion very well, and are used for scientific equipment.

Making glass

Flat glass is made though the float glass process. Molten glass is spread over liquid tin at about 1,832 °F (1,000 °C). The surface of the tin is flat, so the glass becomes flat, too. The molten glass is moved along over the surface of the tin and gradually cooled. When it gets to about 1,112 °F (600 °C), it is hard enough to be lifted out and cut to size.

Compounds of tin

Tin (IV) oxide is added to glass to make it tougher. Thick layers of tin (IV) oxide conduct electricity when sprayed onto glass, and are used to make deicing panels in car windows. Tin (II) chloride is used as a mordant, a chemical that helps a dye stick to cloth.

Compounds containing tin, carbon, and other **elements** are called organotins. Some are very poisonous, such as tributyl tin, a useful ingredient in wood preservatives, **fungicides,** and paints for the hulls of ships. Others are not poisonous, and some are added to polyvinyl chloride (PVC) to improve its properties.

CHAPTER 7

Lead

Lead is a very soft, blue-white metal. Although graphite is sometimes called black lead, it contains carbon, not lead. White lead is actually a range of lead **compounds** that are white in color. Lead often looks dull gray because its surface **reacts** quickly with oxygen from the air to form a thin layer of lead (II) oxide. This layer protects the metal below from reacting with any more oxygen, unless the lead is heated above 1,112 °F (600 °C). It also keeps lead from reacting with water or sulfuric acid. For this reason, lead has been used in the past to make water pipes—some ancient Roman pipes even work today! Lead and its compounds are very poisonous, leading to problems when producing and using it. Lead has a high density, so even small pieces feel heavy.

Each year, 5 million tons of lead are produced worldwide. In its native form, lead is not found as a free **element,** but it is easy to extract from its **ore,** and has been known for thousands of years. The main ore is galena, or lead (II) sulfide. This is mined in many countries, with the majority of production taking place in Australia and China.

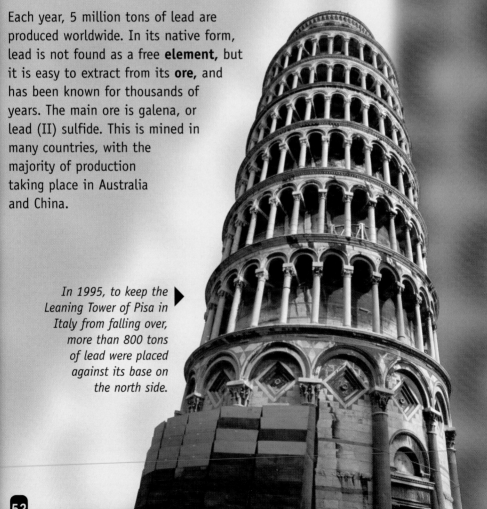

In 1995, to keep the Leaning Tower of Pisa in Italy from falling over, more than 800 tons of lead were placed against its base on the north side.

Lead smelting

Galena is crushed to a fine powder, then treated by a process called flotation separation to remove any waste rock. The powder is mixed with water and special chemicals. Lots of air is bubbled quickly through the mixture to make a froth that floats on the surface, just like a rocky milkshake! The unwanted rock sinks to the bottom, while the froth containing lead sulfide and other metal compounds is piped off.

The froth is dried and mixed with limestone, then heated to about 2,552 °F (1,400 °C) in a limited supply of air. This removes the sulfur from the lead (II) sulfide and turns it into sulfur dioxide.

The equation for roasting lead ore is:

lead (II) sulfide + oxygen → lead oxide + sulfur dioxide
$$2PbS(s) + 3O_2(g) \rightarrow 2PbO(s) + 2SO_2(g)$$

*The sulfur dioxide is used to make sulfuric acid, a substance that is very important for making **fertilizers** and explosives.*

Lead (II) oxide is produced at this stage, but it is impure and it forms solid lumps called sinter. The sinter is crushed and heated with carbon monoxide in a furnace to reach about 2,192 °F (1,200 °C). This removes the oxygen from the lead (II) oxide to make molten lead.

*The equation for **reducing** lead (II) oxide is:*

lead (II) oxide + carbon monoxide → lead + carbon dioxide
$$PbO(s) + CO(g) \rightarrow Pb(l) + CO_2(g)$$

The carbon monoxide is made by heating coke in a blast of hot air.

Lead from the furnace is more than 95 percent pure and is called base bullion. This is purified by a number of complex processes to make lead ready for sale.

Lead the Poisoner

Although lead has many uses, there is a problem with it. Lead is very poisonous, and even tiny amounts of it can harm us. It causes very bad stomach pains if it is breathed in or swallowed. Victims become pale and moody, and later their nerves become damaged, causing paralysis. Children can suffer brain damage and may become deaf or blind. People can die from lead poisoning, so in the modern world strict laws control the use of lead to protect us. In the past, however, this was not the case.

Sugar of lead

Lead acetate tastes sweet, so it is often called sugar of lead. The ancient Romans did not have sugar but liked sweet food and drink, so they used sugar of lead as a sweetener. They made this by leaving sour wine or concentrated grape juice in lead containers. Acetic acid in the sour wine or grape juice **reacted** with the lead to make lead acetate. Unfortunately, Romans who liked sweet food suffered from lead poisoning as a result.

Cider sickness

Cider is a drink made by **fermenting** apple juice. During the 1700s, cider drinkers in one part of England began to suffer from stomach pains, constipation, and other unpleasant symptoms. People thought that sour apples caused the illness, but in 1769 a doctor named George Baker found that it was actually caused by the lead-lined presses and containers used to make the cider. The drinkers were being poisoned because the cider contained large amounts of lead. Once the lead was removed from the equipment used to make cider, the illness disappeared.

Glossy paint

Glossy white paint used to contain lead **compounds** such as lead silicate and lead carbonate to make the color look pure. If the paint was worn away or peeled off, it became a potential health hazard. Young children were particularly at risk because they like to put objects in their mouths. They suffered more from lead poisoning than adults. Modern paints do not contain added lead, but lead poisoning is still a problem in old homes with paint that might contain lead.

Leaded gasoline

When the gasoline in a car engine burns too quickly, the engine rattles or "knocks." Antiknocking substances slow down the burning and help the engine run better. The most common antiknocking substance in the last century was tetraethyl lead, which was first used in gasoline in 1923. Unfortunately, the fumes from the exhaust pipes contained lead, and over 90 percent of the lead in the air in cities used to come from leaded gasoline. Children in particular were poisoned by the lead, and it also damaged the catalytic converters in cars. As a result, many countries have now banned leaded gasoline. The United States introduced unleaded gasoline in 1976, and by 1990 most gasoline sold was unleaded. The United Kingdom banned leaded gasoline in 2000, and now it can be used only in old classic cars that cannot run on unleaded gasoline.

◀ *Homes built before the mid–1900s were often decorated with glossy paint containing lead* **pigments.** *There is a bigger chance of being poisoned by the lead if the paint is peeling off, as seen here.*

Lead in the Modern World

Lead does not **react** with water, so it is used as a roofing material and as a protective covering for underground electric cables. Lead is used in solders to join electronic components together, and in some colored glazes for **ceramics.** The biggest single use for lead is in the manufacture of batteries for cars and other vehicles.

Strict laws control the use of lead to make sure we are not poisoned by it. Modern products that meet the regulations are unlikely to cause lead poisoning because the lead is sealed inside plastic or glass. However, tiny amounts of lead may still escape, so children and pregnant women are usually advised not to eat or drink from lead-glazed crockery. Hot liquids, and acidic foods such as orange juice and wine, cause more lead to escape from their containers. For this reason, wine should not be stored in lead crystal glassware.

Car batteries

Gaston Planté invented the lead storage battery in 1860. This is the rechargeable battery used in many vehicles, including cars, trucks, forklifts, and golf carts. Lead storage batteries are filled with 33 percent sulfuric acid, and they contain plates made of lead and lead (II) oxide. Nearly 80 percent of the lead produced is used to make these batteries, but over 1 million tons of battery lead is recycled each year in the United States alone!

A dense metal

Lead is a dense metal, meaning that even small objects made from lead are heavy. One liter (about one quart) of milk has a mass of about 1 kg (2.2 lb), but the same volume of lead has a mass of 11.34 kg (25 lb)! Divers find it very difficult to stay underwater comfortably without lead weights, and people who fish use lead weights to make their fishing lines sink. However, any lead left behind contaminates the water and may be swallowed by birds. To keep wildlife from being poisoned, many places allow only lead-free fishing tackle. Lead is used to make shotgun pellets and bullets, but lead poisoning is rarely a problem in this use of the metal.

▼ *This patient is being X-rayed. Lead is a very dense metal that will stop radiation. Lead aprons protect patients and staff from getting too much radiation during X-ray imaging. You can even get lead-lined gloves and underwear.*

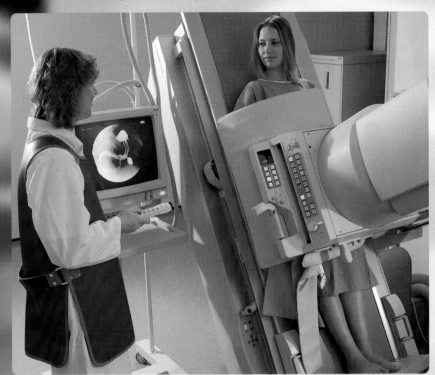

Lead absorbs radiation very well, so the containers used to transport and store radioactive chemicals are lined with it. Equipment that makes radiation, such as X-ray equipment in hospitals, is shielded by lead. Hospital radiographers and other people who use radiation in their work wear aprons containing lead to protect themselves from the harmful effects of radiation.

Glass containing over 65 percent lead oxide is also used for radiation shielding, with the added advantage that you can see through it! Television tubes produce small amounts of X-rays in normal use, so television sets, computer screens, and radar displays use leaded glass to stop the X-rays from leaking. Lead crystal glass, containing about 24 percent lead oxide, is used to make drinking glasses.

Find Out More About the Group 14 Elements

The table below contains some information about the properties of the **elements** in group 14.

Element	Symbol	Atomic number	Melting point °F/°C	Boiling point °F/°C	Density oz/in.³ / g/cm³
carbon	C	6	6,381/3,527	7,281/4,027	1.31/2.27 (graphite) 2.03/3.51 (diamond)
silicon	Si	14	2,577/1,414	5,252/2,900	1.35/2.33
germanium	Ge	32	1,720/938	5,108/2,820	3.08/5.32
tin	Sn	50	450/232	4,716/ 2,602	4.23/7.31
lead	Pb	82	621/327	3,180/1,749	6.56/11.34

Compounds

These tables show you the chemical formulas of most of the **compounds** mentioned in the book. For example, carbon dioxide has the formula CO_2. This means it is made from one carbon **atom** and two oxygen atoms, joined together by chemical **bonds**. All the compounds are solids, except for those on the lists of liquids, gases, acids, and bases.

Oxides

Oxides	formula
aluminum oxide	Al_2O_3
calcium oxide	CaO
germanium (IV) oxide	GeO_2
iron (III) oxide	Fe_2O_3
lead (II) oxide	PbO
silicon dioxide	SiO_2
sodium oxide	Na_2O
tin (IV) oxide (cassiterite)	SnO_2

Carbonates

Carbonates	formula
calcium carbonate	$CaCO_3$
magnesium carbonate	$MgCO_3$
sodium carbonate	Na_2CO_3

Silicon compounds	formula
calcium silicate	$CaSiO_3$
magnesium silicate	$MgSiO_3$
silicon carbide	SiC
sodium silicate	Na_2SiO_3
zirconium silicate	$ZrSiO_4$
silicon dioxide	SiO_2

Germanium compounds	formula
argyrodite	Ag_8GeS_6
bismuth germanate	$Bi_4Ge_3O_{12}$
germanite	$Cu_{26}Fe_4Ge_4S_{32}$
germanium (IV) oxide	GeO_2
magnesium germanate	Mg_2GeO_4

Tin compounds	formula
tin (II) chloride	$SnCl_2$
tin (II) sulfate	$SnSO_4$
tributyl tin	$C_{12}H_{28}Sn$

Lead compounds	formula
lead (II) acetate	$(CH_3COO)_2Pb$
lead (II) oxide	PbO
lead (II) sulfate	$PbSO_4$
lead (II) sulfide	PbS

Acids	formula
carbonic acid	H_2CO_3
hydrochloric acid	HCl
sulfuric acid	H_2SO_4

Find Out More (continued)

Bases

Bases	formula
limewater	$Ca(OH)_2$
sodium hydroxide	$NaOH$

Gases

Gases	formula
carbon dioxide	CO_2
carbon monoxide	CO
methane	CH_4
ethene	C_2H_4
propene	C_3H_6
hydrogen	H_2
oxygen	O_2

Liquids

Liquids	formula
ethanol	C_2H_5OH
silicon tetrachloride	$SiCl_4$
water	H_2O

Other compounds

Other compounds	formula
ammonium sulfate	$(NH_4)_2SO_4$
copper (II) sulfate	$CuSO_4$
glucose	$C_6H_{12}O_6$

Glossary

alkene hydrocarbon with two or more carbon atoms joined to each other by double bonds (two bonds, not just one)

allotrope one of two or more different forms of an element. Allotropes have the same chemical properties but different physical properties.

alloy mixture of two or more metals, or mixture of a metal and a nonmetal.

amorphous not containing any crystals

anodizing making a layer of oxide on the surface of a metal using electrolysis

antioxidant substance that prevents oxygen from reacting with other chemicals

atom smallest particle of an element that has the properties of that element.

atomic number number of protons in the nucleus of an atom

bond force that joins atoms together

brittle likely to break into small pieces

bunny suit protective clothing that stops dust, skin, and hair from getting into a "clean room"

carbohydrate compound that contains carbon, hydrogen, and oxygen atoms. Sugars, starch, and cellulose are carbohydrates.

catalyst substance that speeds up reactions without getting used up

ceramic tough solid made by heating clay and other substances to high temperatures in an oven

combustion chemical reaction that produces heat, usually by a fuel reacting with oxygen in the air

compound substance made from the atoms of two or more elements, joined together by chemical bonds

destructive distillation method used to produce new substances by heating solid or liquid carbon compounds to a very high temperature minus oxygen

distillation method used to separate a liquid from a mixture of a liquid and a solid

electrolysis process of breaking down or decomposing a compound by passing electricity through it. The compound must be molten or dissolved in a liquid for electrolysis to work.

electron particle in an atom that has a negative electric charge. Electrons are found in shells around the nucleus of an atom.

element substance made from only one type of atom

enzyme substance made by living things that controls the chemical reactions that happen inside the living thing

fermentation reaction caused by the enzymes in tiny fungi called yeast. In fermentation, sugar is broken down to make alcohol and carbon dioxide.

fertilizer chemical that gives plants the elements they need for healthy growth

fossil fuel fuel made from the ancient remains of dead animals or plants. Coal, oil, and natural gas are fossil fuels.

fractional distillation type of distillation that is used to separate mixtures of two or more liquids

fungicide chemical that kills the fungus that can damage crops

group vertical column of elements in the periodic table. Elements in a group have similar properties.

hydrocarbon compound made from hydrogen and carbon atoms only

impermeable rock rock that does not let liquids and gases through it

insecticide chemical that kills insects that can damage crops

malleable able to be bent into shape without breaking. Metals and alloys are malleable.

mineral substance that is found naturally in the earth but does not come from animals or plants. Metal ores and limestone are examples of minerals.

neutralization reaction between an acid and an alkali or a base. The solution made is neutral, meaning that it is not acidic or alkaline.

neutron particle in an atom that has no electric charge. Neutrons are found in the nucleus of an atom.

nucleus center part of an atom made from protons and neutrons. It has a positive electric charge.

ore substance containing minerals from which metals can be taken out and purified

period horizontal row of elements in the periodic table

periodic table chart in which all the known elements are arranged into groups and periods

permeable rock rock that lets liquids and gases through it because it contains tiny holes and cracks

photolithography process of transferring shapes from a mask to the surface of a silicon wafer

photosynthesis chemical reaction in which green plants make sugars from carbon dioxide and water, using the energy from light. Oxygen is also made.

pigment solid substance that gives color to a paint or other substance. A pigment does not dissolve in water.

polymer large molecule made from lots of smaller molecules (monomers) joined together

product substance made in a chemical reaction

proton particle in an atom's nucleus that has a positive electric charge

reaction chemical change that produces new substances

reduction removal of oxygen from an element or compound in a chemical reaction, or addition of electrons

refining removing impurities from a substance to make it more pure. It can also mean separating the different substances in a mixture, for example, in oil refining.

semiconductor substance, such as silicon, that is an electrical insulator at room temperature, but a conductor when it is warmed or other elements are added to it

zone refining method used to make pure crystals by melting and freezing impure crystals

Timeline

carbon discovered	ancient times	unknown
tin discovered	ancient times	unknown
lead discovered	ancient times	unknown
diamonds are made of carbon	1772	Antoine Lavoisier
graphite is made of carbon	1779	Carl Scheele
silicon discovered	1823	Jöns Berzelius
germanium discovered	1886	Clemens Winkler
buckminsterfullerene, C_{60}, discovered	1985	Richard Smalley, Robert Curl, and Harold Kroto
ununquadium first made	1998	Lawrence Berkeley National Laboratory

Further Reading and Useful Websites

Books

Fullick, Ann. *Science Topics: Chemicals in Action*. Chicago: Heinemann Library, 1999.

Oxlade, Chris. *Chemicals in Action* series, particularly *Atoms; Elements and Compounds*. Chicago: Heinemann Library, 2002.

Websites

WebElements™
http://www.webelements.com
An interactive periodic table crammed with information and photographs.

DiscoverySchool
http://school.discovery.com/students
Help for science projects and homework, and free science clip art.

Proton Don
http://www.funbrain.com/periodic
The fun periodic table quiz!

Mineralogy Database
http://www.webmineral.com
Lots of useful information about minerals, including color photographs and information about their chemistry.

Index